LEARNING JOURNEYS

written by Eunice M. Magos
and Esther H. Hornnes

illustrated by Priscilla Burris

EUNICE M. MAGOS received a Bachelor of Science degree from New York University, and a Master's degree in Individualized Education from the College of St. Scholastica in Minnesota. She has been a director of Head Start, taught gifted primary classes and participated in the Learning to Read Through the Arts program. She has taught remedial reading and kindergarten, and currently teaches first grade in the Hopatcong Borough School District in New Jersey.

ESTHER H. HORNNES received a Bachelor of Arts degree from Shelton College and William Paterson College in New Jersey. She has done graduate work at North Dakota State University and Oslo University in Oslo, Norway. She has taught grades 1-4 and presently teaches pre-school in the Hillside Nursery School in Succasunna, New Jersey.

PRISCILLA BURRIS received an Associate of Arts degree in Creative Design from the Fashion Institute of Design and Merchandising in Los Angeles. As a free lance artist of child-related artwork, she has been drawing since she was one year old. Priscilla lives in southern California.

Reproduction of these pages by the classroom teacher for use in the classroom and not for commercial use is permissible. Reproduction of these pages for an entire school or school system is strictly prohibited.

Copyright 1985 by THE MONKEY SISTERS, INC.
22971 Via Cruz
Laguna Niguel, CA 92677

ISBN 0-933606-35-4

LEARNING JOURNEYS

Learning Journeys presents interesting and informative lessons about the people, foods, customs, games and crafts of other lands while children learn skills in reading, writing, math, and research.

Each set of *READING* skill pages includes a story page followed by six questions for children to answer. Writing space is included on the page.

There is a page listing unusual or little-known facts about the people and country along with a research activity for the students.

The *WRITING* activity contains information about a food associated with the country and directions to complete a motivating task relating the writing activity to the food.

Each set of *MATH* skill pages includes background or directions about a game or craft particular to the country plus a math drill and practice page in a motivating format.

When the children complete the unit on a country, they will have worked on improving skills in reading, research, writing, and 'rithmetic while learning about life in other parts of the world.

The countries may be presented in any order to coordinate with your curriculum. **Learning Journeys** is a two-book series as follows:

BOOK 1	BOOK 2
England	France
Holland	Germany
Switzerland	Italy
Israel	USSR
Puerto Rico	India
Japan	Australia
Canada	Mexico
Saudi Arabia	Nigeria

Contents Pages

	Reading	Comprehension	Did You Know?	Writing	Game/Craft	Math
France	1	2	3	4	5	6
Germany	7	8	9	10	11	12
Italy	13	14	15	16	17	18
USSR	19	20	21	22	23	24
India	25	26	27	28	29	30
Australia	31	32	33	34	35	36
Mexico	37	38	39	40	41	42
Nigeria	43	44	45	46	47	48

Learning Journeys Passport

This passport is issued to

_____

and is valid in the following places:

- France ☐
- Germany ☐
- Italy ☐
- USSR ☐
- India ☐
- Australia ☐
- Mexico ☐
- Nigeria ☐

Each box may be checked when you have completed the unit for each place.

MEET A FRIEND FROM FRANCE

"Bonjour!" to you from the French Republic. I am Jacqueline, and I bring you greetings from my country. We are a country that is well-loved by its people. We say La Belle (beautiful) France, and display our tri-color flag of red, white, and blue everywhere. Our people can differ greatly and resemble dark, olive-skinned Mediterraneans or tall, blond, blue-eyed Scandinavians. But we are alike in our language. French can be recognized the world over. As we talk, we gesture broadly with our hands, and we have very expressive faces. The city people try to dress fashionably and we are famous for our fine cooking.

Napolean Bonaparte, Joan of Arc, Jules Verne, Louis Pasteur, Charles de Gaulle, and Jacques Cousteau are the names of some well-known French men and women.

France, smaller than Texas, is the second largest country in Europe, after European Russia. There are 2,000 miles (3,219 km) of coastline, and a moderate climate well suited to farming. Apples, and grass for dairy farms that produce rich milk for cheese and butter, grow in Normandy. Further south, wheat fields are seen, as well as grapes and olives. In the southern sections, lemons, palm trees, and sub-tropical flowers for perfume grow. To the east in the foothills of the Alps, you will see yellow mustard fields. The town of Dijon gives its name to a famous mustard.

In the mountain regions of the Pyrenees, Massif Central, and Alps, the winters are very cold and snowy, but are enjoyed by skiing enthusiasts. Along the Mediterranean Sea coast, called the Riviera, are the world famous beach resorts of Nice and Cannes.

I live in our capital city, Paris. It is one of the most beautiful cities in the world. Notre Dame Cathedral, the Bastille, the Eiffel Tower, the Louvre Museum, and the Arch of Triumph are famous places to see. Along the banks of the river Seine you can see painters along with book or flower sellers. There are narrow winding streets with quaint stone buildings and wide boulevards with outdoor cafes.

We live in a small apartment building just as many other French people do. Wooden window shutters keep out the glare of the sun and street noises. New supermarkets are being built, but my mother likes to shop the old-fashioned way by going to a different store for groceries, cheese, bread, candy, dairy, fish, or meat. Local outdoor markets sell a variety of foods weighed in kilograms and grams.

My father drives his car to work at a huge automobile factory just outside Paris. Hydro electric and nuclear power have helped us become a modern industrial nation.

We study hard at school, and have a long day with few vacation days during the school year. There isn't much time for sports, but we do watch TV. My brother likes to watch famous French skiers and a 20-day cycle race called the Tour de France. I look forward to our 10-week summer vacation when I will be going to camp to enjoy hiking, cooking, hobbies, and sports. Next year my class will be going to the Alps combining a week of school with ski instruction.

Bastille Day, our national holiday, is my favorite. It is July 14. We have fireworks, parades, and parties to celebrate our democratic system of government. When we sing our national song, La Marseillaise, it reminds me how much I love France!

Skill: Reading comprehension

Name _____

Date _____

MEET A FRIEND FROM FRANCE

Answer these questions after you have read about France.

1. Name three ways that the French express themselves.

   _____
   _____
   _____

2. Describe the city of Paris.

   _____
   _____
   _____

3. Do research to find out more about two well-known French people mentioned in the reading. Tell why these people became famous.

   _____
   _____
   _____

4. What products are grown in France?

   _____
   _____
   _____

5. Describe the old-fashioned way of shopping for food in France.

   _____
   _____
   _____

6. Compare Bastille Day to one of your national holidays. How are they alike?

   _____
   _____

Learning Journeys © THE MONKEY SISTERS, INC.

Skill: Researching

FACTS FROM FRANCE

Did you know that...

- The world's longest road tunnel, 7 miles (11.2 km) runs under Mont Blanc, the highest peak in the Alps, and links France with Italy.

- France is thought of as the star-shaped country.

- The Statue of Liberty in New York Harbor was a gift from the French people to the American people symbolizing the friendship between the two countries.

- The Eiffel Tower is 984 ft. (300 m) high, has 2,710 steps, and the view from the top is 30 miles (48.3 km) in all directions.

- Telephone numbers are grouped in twos. 93 71 22 would be read as ninety-three, seventy-one, twenty-two instead of nine, three, seven, one, two, two.

Research the names of at least three famous French artists. Tell what type of art each was famous for.

___

___

___

___

Skill: Dividing words into syllables; using accent marks

Name _____

FRENCH BREAD

"Bon Appetit!" means good appetite, and when the French have a meal, they enjoy eating. French chefs are world famous for their fine dishes and artistic combinations of textures, shapes and colors of ingredients. Dining in a French home is important. Having the family together with good conversation and appetizing food make the meal something special.

One staple that is found at almost every meal is the baguette. This is a long, crusty loaf of bread. Each morning, hot loaves are bought from the local bakery to be eaten that day.

BREAD
1 package active dry yeast
2 cups (480 ml) warm water
1 teaspoon (5 ml) salt
2 teaspoons (10 ml) sugar
5-6 cups (1200-1440 ml) white flour

Dissolve one package of yeast in 2 cups (480 ml) warm water. Add salt, sugar, and 3 cups (720 ml) of the flour. Stir until smooth. Then add additional flour until dough is stiff (usually 2-3 cups [480-720 ml] more). Knead on a floured board until smooth and elastic, about 10 minutes. Let rise in a bowl covered with a clean towel until double in size. Punch down and shape into two long loaves. Place loaves on cookie sheet. Cut 3-4 diagonal slashes across loaves. Cover with towel and let rise again until double in size. Sprinkle with water.

Bake at 375° (191° C) for 35-40 minutes. Sprinkle more water on top a couple of times during baking to give the bread its chewy French crust.

Serve warm the way it is, or with a small bowl of soup to dip the bread in as the French people do.

Accent marks are shown in the dictionary to help you pronounce words correctly. The accented syllable is spoken more strongly than the unaccented syllables. Divide these words into syllables and place the accent mark on the correct syllable. Check your work by looking up each word in the dictionary. Example: sprin'kle

appetite _____   combinations _____

textures _____   together _____

ingredients _____ conversation _____

appetizing _____   elastic _____

additional _____   artistic _____

important _____   diagonal _____

Learning Journeys © THE MONKEY SISTERS, INC.

Craft project—France

STRAWBERRY MAGNETS

For a long time, figures and fruits made from Baker's Clay have been popular with the French people. At Christmas time, manger scenes were made from dried clay and then painted. Ornaments in the shapes of baskets, fruits and flowers were also made.

Peddlers from Marseille would sell these clay figurines and clay baskets filled with clay fruits. These became popular year-round. Until the 1800's, the Baker's Clay crafts were dried in the sun. Then firing in an oven called a kiln was used. Craftspeople who want to follow the old traditional ways still let their clay item dry in the air or sun before painting.

BAKER'S CLAY

1 cup (240 ml) salt
2 cups (480 ml) flour
½ teaspoon (3 ml) baking powder
1 cup (240 ml) water

Mix dry ingredients in a bowl. Stir in water. Flour hands and knead mixture 5-10 minutes until it feels like clay.

Flatten out with a rolling pin. Use wax paper on both sides to keep clay from sticking to table top or rolling pin.

Cut a strawberry pattern out of cardboard. (Other fruits can be made instead.) Lay pattern on clay. Cut around berry. Do the same with pattern for leaves.

Let air dry for a couple of days. Glue leaves to berry. Paint. Seal with a coat of clear nail polish. Glue a small piece of magnet strip on the back.

Learning Journeys © THE MONKEY SISTERS, INC.

Skill: Recognizing odd and even numbers   Name _____

Fill in missing numbers where needed. Skip count—odd or even.

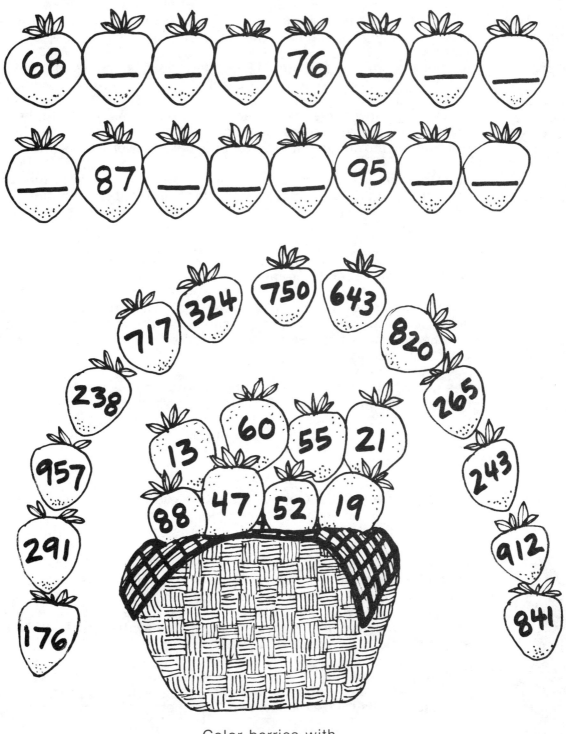

Color berries with odd numbers red; even numbers green.

GREETINGS FROM GERMANY

Guten Morgen (Goo'ten Mor'gen)! That's "Good Morning" in German. I'm Ursula and I live in West Germany. I'd like to introduce you to my country. Germany is located in Europe. The size of Germany is just a little smaller than the state of Montana in the United States.

In the northern part of Germany you find the lowlands which are low and flat. In the central and southern parts of Germany you find rolling hills, plateaus, and mountain ranges, one being the Bavarian Alps. Two famous rivers which flow across the lowlands are the Rhine and the Elbe. You will find cargo ships as well as pleasure boats on the Rhine River. The rivers of Germany provide transportation and water power for manufacturing.

The Germans are creative people. We are known throughout the world for scientific and artistic achievements. One important contribution made by a German was the X-ray. This was discovered in 1895 by Wilhelm Roentgen. Ludwig von Beethoven was one of Germany's greatest composers.

Today Germany is divided into two parts, East Germany and West Germany. They have governments that are very different. East Germany is a dictatorship. The people have to follow orders from a person they did not choose to have as their leader. West Germany is a democracy. The people help decide how their government should be run and who their leader will be. The capital of East Germany is East Berlin and the capital of West Germany is Bonn.

The highways in West Germany are some of the best in Europe. Our super-highways are called autobahns and you can travel for miles without stopping or slowing down. More and more people are buying cars, but many still travel by bus, bicycle or motorcycle.

Most cities in Germany are very old. There are a lot of old buildings. In some cities, however, there have been many changes and you see many new modern buildings. Outside the cities you will see gardens that are well taken care of. These gardens belong to families who live in the city. They go out and take care of their gardens on Sunday afternoons.

The German people enjoy singing and dancing. We enjoy many festivals during the year. I think my favorite festival is the Oktoberfest. It is held in Munich at the fairgrounds. I'm getting a new costume to wear this year and I can hardly wait. We usually watch the horse-riding tournament first, then a parade, and finally we sing and dance in the streets. We also enjoy eating delicious German sausages.

Germany is a beautiful country. Our climate is quite enjoyable.
The temperature is moderate in summer and winter. Maybe you would like to come and visit us someday. We'd like that!

Skill: Reading comprehension

Name _____

Date _____

GREETINGS FROM GERMANY

Answer these questions after you have read about Germany.

1. How does the land of Germany differ in the northern, the central and the southern parts?
   _____
   _____
   _____

2. Why are the rivers in Germany so important? Name two of them.
   _____
   _____
   _____

3. Give two examples of artistic and scientific achievements of the German people.
   _____
   _____
   _____

4. How do the governments of East Germany and West Germany differ?
   _____
   _____
   _____

5. In general, how would you describe the German people?
   _____
   _____
   _____

6. Use your dictionary to find the meaning of the word *festival*.
   _____
   _____

Learning Journeys © THE MONKEY SISTERS, INC.

Skill: Researching   Name _____

JUST ABOUT GERMANY

Did you know that . . .

- Little Red Riding Hood, Hansel and Gretel plus many other fairy tales started in Germany.

- The German people celebrate Christmas from December 1 - January 6.

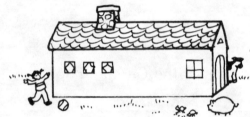

- On a farm in Germany, the house and barn are one building. The animals live in one end and the people in the other end.

- "Danke Schön" means thank you in German.

- Pretzels were made first in Germany.

It is interesting to learn unusual facts about people and places. Read more about Germany and list two unusual facts you learned. Be ready to share them with the class.

- _____
  _____

- _____
  _____

Learning Journeys © THE MONKEY SISTERS, INC.

Skill: Writing sentences in a paragraph.

Name _____

FRANKFURTER FACTS

A frankfurter is meat which is ground up and stuffed into a skin. The German people were the first to invent this sausage. It really started in the city of Frankfurt, Germany. That is why it was given the name frankfurter. The Germans who immigrated to America introduced the people of the United States to the frankfurter. It is better known to some as the "hot dog."

A favorite way to serve the frankfurter in Germany is to simmer it in water on the top of the stove for a few minutes until it is plump and hot. Eat it with sauerkraut, which is shredded cabbage aged in vinegar. Other ways to eat a frankfurter would be to put it in a frankfurter roll and top it with mustard, ketchup, onions or a pickle relish.

When you eat your frankfurter, you will say "hot dog"!

Pretend you are on a camping trip. How would you prepare your frankfurter then? Use complete sentences in a paragraph. Here are some words you may need to use:

| campfire | sticks | grill | barbeque |
| outdoors | crackle | delicious | light |

_____
_____
_____
_____

Learning Journeys © THE MONKEY SISTERS, INC.

10.

Game directions—Germany

NINEPINS

Ninepins is a bowling game played in Germany. A player rolls a ball at nine pins set in a diamond shape. It was so popular in Germany that village dances and celebrations of baptism included bowling.

The nine wooden clubs are called "kegles" and the bowlers are known as "keglers." Bowling in Germany dates back to the Middle Ages. From there, it spread to the rest of Europe.

Teacher directions for game: Ask the children to each bring in a bleach bottle. A quart-size round one works best. Remove the paper label. Number bottles with felt tip marker as shown. Set up two sets of nine bottles each about ten feet apart in a diamond shape (1-2-3-2-1). Divide the class into two teams. Give each team a small rubber ball about seven inches in diameter. Each child will try two times to knock the bottles over by rolling the ball. Add up the number of bottles knocked over in the two tries and write the score down. Continue until ten frames are played. Score as you would for bowling played in the United States.

Arrangement of bowling pins (bleach bottles) on floor.

Learning Journeys © THE MONKEY SISTERS, INC.

Skill: Subtraction with regrouping   Name _____

Complete subtraction problems. Use a RED crayon to color each pin with an answer of nine. What pattern do you see?

"BUON GIORNO" FROM ITALY

"Buon Giorno!" Hello, my name is Lorenzo and I bring you greetings from the Italian Republic, the country shaped like a boot. Italy, as we are called, is the size of New Mexico and has one-fourth the population of the USA.

We are in southern Europe along the Mediterranean Sea, and have a warm climate except for the northern mountain regions that border France, Switzerland, Austria, and Yugoslavia. There you will see snow-covered peaks, and glaciers. Villages in these Alpine areas have stone houses with thick walls to keep out the cold, and steep streets. People travel from around the world to ski in the winter and climb in summer.

The Appenine Mountains run through the length of Italy. These mountains extend into Sicily, our large island at the "toe" of Italy. Another large island that is part of our country is Sardinia at the "knee" of the "boot," across the Tyrrhenian Sea opposite our capital city, Rome.

We have two live volcanoes. Mount Vesuvius is near Naples and Mount Etna is in Sicily.

Our beaches on the northwest coast along the Adriatic Sea are called the Italian Riviera. They are narrow in many places, and are covered with pebbles instead of sand.

The river Po, and the rivers and lakes that feed into it, form a big valley in the northern section. Here in the Po Valley is most of Italy's fertile soil for growing, and its industry.

Our sea ports such as Venice, Genoa, and Trieste are important. We ship cloth and clothing, leather goods made around Florence, cheeses, olive oil, vehicles, and glasswares to many other countries.

Rome, called Roma, is my home. It is a city built on the banks of the Tiber River. Low hills on its eastern banks, where ancient Romans built the city, are called the Seven Hills of Rome.

My father is a tour guide. He drives a taxi and shows visitors the wide boulevards, narrow twisting streets, and famous buildings like the ancient Coliseum. People say that, "All roads lead to Rome," because it is a center for learning about history, art, and architecture.

The most famous Italian artists are Leonardo da Vinci, who was also an inventor and scientist, and Michelangelo. Maria Montessori was a famous educator. Julius Caesar, a Roman leader, established the modern calendar, and has the month of July named for him. Marco Polo, Christopher Columbus, Amerigo Vespucci, John Cabot, and Giovanni da Verranzano were Italians who explored the world 500 to 600 years ago.

I go to school and study the same subjects that you do. Our dress is much the same as other western children. I intend to go to senior high, and then to a technical school to learn a trade. Each day after school my friends and I play soccer at a park. It is our favorite sport along with football.

We are a large family with brothers, sisters, parents, and grandparents all living at my house. Everyone eats together at dinner time, and shares the events of their day. It can get quite noisy sometimes! As part of our meal, pasta is always served. The best known pasta is spaghetti.

I am looking forward to Carnival time. That is six weeks before Easter. Cities and towns have parades with flower decorated floats, and costumed figures. It is a lot of fun!

We would like to see you then, or anytime. Let's go skiing, or visit a fishing village along the sea coast. You will enjoy yourself, I'm sure!

Skill: Reading comprehension

Name _____

Date _____

"BUON GIORNO" FROM ITALY

Answer these questions after you have read about Italy.

1. Find Italy on a map. Tell why it is compared to a boot and parts of a leg. Include the names of these comparisons in your answer.

   _____
   _____
   _____

2. How are the beaches of the Italian Riviera different from most seaside resorts? Would you like to sun bathe there? Tell why.

   _____
   _____
   _____

3. Use your dictionary to find the meaning of the word *coliseum*. What is a coliseum? Compare the ancient Coliseum in Rome to something similar in your country.

   _____
   _____
   _____
   _____

4. Name two things for which Julius Caesar is known.

   _____
   _____
   _____

5. Explain the meaning of the saying, "All roads lead to Rome."

   _____
   _____
   _____

6. Look around your house or school for products made in Italy. Make a list of these products. Which two do you like the best? Tell why.

   _____
   _____
   _____

Learning Journeys © THE MONKEY SISTERS, INC.

INFORMATION ABOUT ITALY

Did you know that . . .

- The Leaning Tower of Pisa has slowly sunk on one side, but visitors may still walk the steps to the observation rail at the top.

- Italian author Carlo Collodi wrote the story of *Pinocchio*, the puppet who wanted to be a boy.

- The Appian Way was constructed in 312 B.C., and is still a major road into Rome.

- In Venice, a city of canal streets, traffic moves by boat. There are policeboats, fireboats, taxiboats, powerboats, and the most famous of them all, the gondolas.

- Almost every town or city has at least one *piazza* or village square where people come to meet and shop because Italians love to get together in crowds.

Rome and Venice are two well-known Italian cities. Using a map, find five more well-known cities. Next to each city, list one reason for the city's fame.

_____

_____

_____

_____

_____

Skill: Using nouns and verbs; humorous writing.   Name _____

ITALIAN ICES

Some people say that Italy has the best ice cream in the world. Just ask Italians and they will agree!

Anytime is an excuse for Italians, who love ice cream, to look for *gelato* (ice cream in Italian). After a stroll along the market place or after an outing in the park the sweet, cool cream or ice is a welcome treat.

Gelati is an ice cream made from milk and is very much like the flavors in Canada and the USA.

Granita is a sherbet made with ice and sweet syrup.

A natural, no sugar *Italian Ice* or Granita can be made with any unsweetened fruit juice. Lemon and coffee are popular flavors in Italy, but would require the addition of sugar.

Italian Ice on a Stick

Unsweetened, undiluted fruit juice is poured into wax-coated, small paper cups. Insert clean popsicle sticks into the cups when the juice is partially frozen. That way sticks will stand in the center of the cups. Keep in freezer until the entire cup of juice is frozen. When entirely frozen, peel cup away and begin to lick and enjoy!

Suggested juices are: pineapple, orange, grape, apple, apricot, or any other natural juice.

Pick a verb to rhyme with each noun given. Then write a silly sentence using each pair. Example: Noun—ice cream
Verb—dream.

"I can dream that I'm swimming in ice cream."

	NOUN	VERB
1.	stick	_____
2.	park	_____
3.	treat	_____
4.	market place	_____

1. _____
2. _____
3. _____
4. _____

Learning Journeys © THE MONKEY SISTERS, INC.

Craft project—Italy

RUBBERBAND VIOLINS

Italians love music. You will find them singing everywhere.

Strolling musicians are popular in restaurants. Waiters and patrons will join the musician in song as he plays his violin. Songs may be the latest popular numbers or the Italian favorite, opera!

Italians will find an excuse to sing about anything. The famous song "Funiculi, Funiculà" is about the Finicular Railway built up the side of Mount Vesuvius.

Many words in music are Italian. *Solo* means alone. *Largo* is slow. *Allegro* is fast. *Forte* is loud, and *fortissimo* is very loud. *Pianissimo* means very soft in Italian.

Piano, piccolo, cello, and violin are Italian names of instruments.

VIOLINS

You will need an assortment of rubber bands, an unsharpened pencil, and a box with an opening in the top such as a tissue box. However, any size box that isn't larger than a tissue box will work nicely.

Pull one rubber band, that is as wide as the pencil eraser, from end to end onto the pencil. This is your bow.

Stretch four rubber bands around the box so that they span the opening and are at least one-half inch apart.

Experiment with various widths of rubber bands to produce the sound you like best when the bow is run across the strings (bands) of your violin.

Find an Italian record such as *O Sole Mio* sung by the great Italian tenor Enrico Caruso to play as you accompany the music with your "violin."

Violin

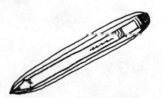
Pencil bow with rubber band covering from end to end.

Rubber bands strung crosswise or lengthwise over box.

Skill: Addition/subtraction of hundreds with regrouping   Name _____

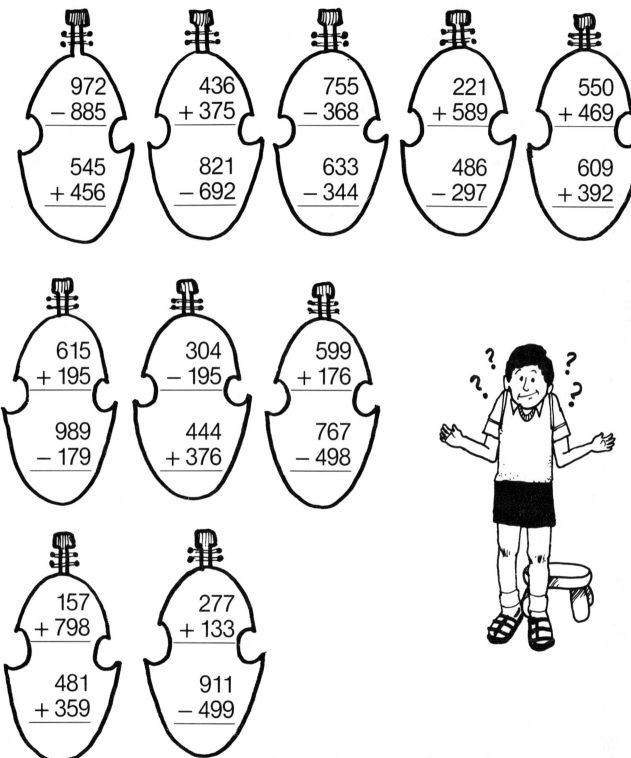

972 − 885
545 + 456

436 + 375
821 − 692

755 − 368
633 − 344

221 + 589
486 − 297

550 + 469
609 + 392

615 + 195
989 − 179

304 − 195
444 + 376

599 + 176
767 − 498

157 + 798
481 + 359

277 + 133
911 − 499

Complete each set of examples.
Then find Mario's violin.
It's the violin where both answers
are the same. Color it brown.

READ ABOUT RUSSIA

"Doh-broh-yeh Oo-troh!" My name is Olga and I bring you greetings from the Union of Soviet Socialists Republics . . . or Russia.

The USSR is the world's largest country covering one-sixth of the earth's land area on two continents, Europe and Asia. It is 2-1/2 times the size of the U.S. and includes 11 time zones with a 10-hour difference from the Pacific Ocean in the east to Poland in the west. The Arctic Ocean borders in the north, and Turkey, Mongolia, and China border to the south.

Our climate ranges from very cold in Arctic Siberian regions with perma frost to mild temperatures in the south along the Black Sea where vacationers like to go. But most of the USSR has long, bitter winters with low temperatures, frozen lakes and rivers, and heavy snow. Our *targa* pine forests are the largest in the world. We are second to the U.S. in agricultural products and second to Japan in fishing. We have many minerals such as iron ore, manganesze, copper, silver and gold, and we lead in the world's production of oil and coal.

The official language is Russian, but there are over 100 national groups speaking over 150 languages. Eskimos are in northern Siberia, Ukrainians are farmers and miners, and the Georgian and Armenian people are famous for their folk dances.

I live in Moscow, the capital of Russia, one of our 15 Republics. This is where the Kremlin, the seat of government for all of the USSR, is located. We were once ruled by tsars (kings), but now one political party rules and they choose the leader. Red Square (this means *Beautiful Square* in Russian) is next to the Kremlin. Here we have parades in May and November. School children, soldiers, athletes, workers, floats, and military vehicles parade past the tomb of the founder of modern Russia, Vladimir Lenin. Our beautiful and cultural city, Leningrad, was named for him.

Our flag is red with a hammer and sickle. That stands for the workers and farm peasants. We call each other, "Comrade" to show that all are equal.

My mother, father, brother, *babushka,* which means grandmother, and I live in a very small apartment. In the living room, we have a dining table, TV, cabinets, potted plants, and divans for grandmother and the children to sleep on. Some important people are lucky enough to have a *dacha* (country cottage) for vacations away from the city.

We don't own a car, but we travel by bus or metro. That's our decorated underground subway which even has lighted chandeliers in the passageways. Shopping is interesting at GUM department store. It is like an indoor mall with a huge glass roof and fountains. I also like the Moscow Circus and Gorky Park Amusement Center in the summer.

I go to school six days a week from 8:00 to 1:00. After lunch, we participate in clubs and activities like Young Pioneers, sports, music, art, and social or political activities. I study ballet and dream of dancing one day with the Bolshoi Ballet Company. Russians are known for world champion chess players, Olympic ice skaters, and gymnasts like Olga Korbut.

Some famous Russian writers are Chekov, Pasternak, and Tolstoy. Our famous composers are Rachmaninov, Rimsky-Korsakov, Stravinsky, Tchaikovsky, and Prokofiev who wrote "Peter and the Wolf."

I like Russian music very much, but do enjoy your "pop" music and dances. If you ever visit, perhaps you could show me some new steps!

Skill: Reading comprehension

Name _____

Date _____

READ ABOUT RUSSIA

Answer these questions after you have read about the USSR.

1. Leningrad is in the most western time zone. If it is 7 a.m. there, what time would it be on the Pacific coast in the farthest time zone? Compare times for 12 noon, 4 p.m. and 8 p.m.

   _____
   _____
   _____

2. What are some of the different national groups in the USSR? How do they differ?

   _____
   _____
   _____

3. What is the climate like in the USSR? How do you think the people dress?

   _____
   _____
   _____

4. Who is Vladimir Lenin? In what way is he remembered?

   _____
   _____
   _____

5. Give the meaning of babushka and dacha. Use complete sentences.

   _____
   _____
   _____

6. Pick a famous Russian composer to research. Write about his life and music. Choose a recording of one of his compositions to play for the class.

   _____
   _____

Learning Journeys © THE MONKEY SISTERS, INC.

Skill: Researching

READ ABOUT RUSSIA

Did you know that . . .

- Russian words are written in the Cyrillic alphabet. CCCP written on the tails of Russia's Aeroflot Airline's jets, or on shirts of Russian ice hockey players, means the same as USSR in our Roman alphabet.

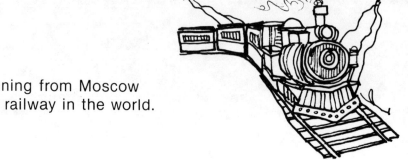

- The Trans-Siberian Railway running from Moscow to Vladivastock is the longest railway in the world.

- During the winter in Irkutsk you can buy your milk frozen with a stick in the middle. Children use the stick to carry it home and let it thaw out in a container.

- Over 70% of the doctors in the USSR are women.

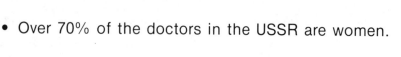

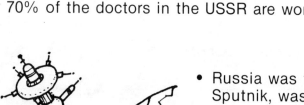

- Russia was first in space. The first space satellite, Sputnik, was launched in 1957, and in 1961 Yuri Gagarin was the first person to orbit the earth.

Make a copy of the Russian, or Cyrillic alphabet. Then write your name in this alphabet.

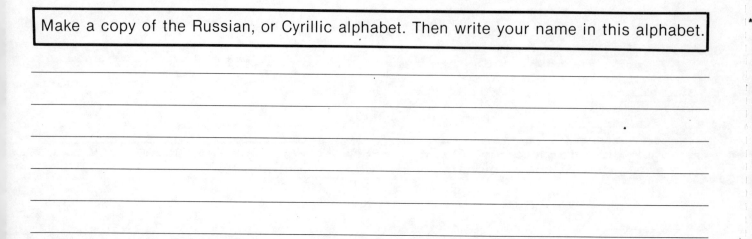

Skill: Creative writing

BEET SOUP IS HARD TO BEAT!

Russian food is known to be delicious and filling. At breakfast time during the winter school children might have a hot bowl of *kasha* oatmeal porridge.

Cheese or salami is eaten on a dark brown rye bread called "black bread" by the Russians. On special occasions pancakes called *bliny* are eaten with sour cream, jam, herring or special fish eggs called *caviar* which are an expensive luxury.

Russians drink tea at all meals. It is made in a special metal *samovar* urn that has a small charcoal fire in the base to heat the water.

Supper is often vegetable soup. A favorite is *borsch* made from cabbage and beetroots. The beets give the soup its red color. It can be served hot or cold, but is tasty and warming, served hot topped with sour cream and eaten with dark bread.

BORSCH

4 cups (960 ml) water or beef broth
2 carrots
1 onion
2 stalks celery

2 large beets
½ head of a very small cabbage
Salt, pepper, bay leaf to taste

Slice all vegetables and save ½ beet and cabbage. Shred these. Cover and cook in large pot until tender. A teaspoon (5 ml) of sugar and a tablespoon (15 ml) of vinegar can be added for zip 10 minutes before serving.

Serve in paper bowls or use hot cups. Top each individual serving with sour cream.

Beet is a vegetable. A homonym for this word is beat. Use your imagination and include both words in a silly story about "The Beet that Beat-it!" Try to include some dialogue. Use quotation marks correctly.

_____
_____
_____
_____
_____

Learning Journeys © THE MONKEY SISTERS, INC.

Game directions—USSR

FIELD DAY

Sports are extremely important in the USSR. Many sports clubs are organized throughout the country. Participation in physical culture is required of school children from kindergarten through college.

There are junior sports schools for children ages 9 to 14, and a nationwide competition is held every two years. Special ribbons and certificates are awarded to these young athletes.

Many go on to the USSR Games, which are held every four years. These winners will go to international and Olympic competitions. Track and field events are one of the most popular areas of sports competition in the USSR.

A *Field Day* can be organized with a variety of competitions and prove to be an exciting event. Another class or classes in your school could be invited to make the day or class period a real event.

Suggested events might include:
- Relay race
- Dash
  a. Boys
  b. Girls
- Long distance run
- "Discus-type" event using a frisbee
- Throwing event with bean bag or ball
- Standing broad jump

Boys and girls can be on the same teams if there is an even number of each.

Decide in advance the location of each event. If there are a lot of children involved, then do more than one event at the same time. Invite parents to cheer on teams and to help out.

At the conclusion of the Field Day, award paper blue ribbons to all participants.

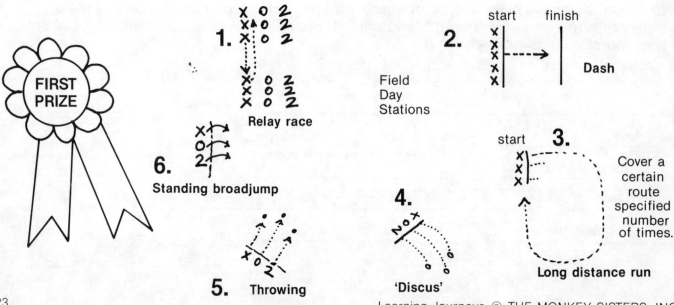

Field Day Stations

1. Relay race
2. Dash
3. Long distance run — Cover a certain route specified number of times.
4. 'Discus'
5. Throwing
6. Standing broadjump

Skill: Metric measurement with centimeters and millimeters   Name _____

USSR

Measure the left side of each ribbon using a metric ruler.

INTERESTING INDIA

"Ram-ram!" That's hello in Hindi. My name is Kittu and I live in the country of India. India is about one-third the size of the United States.

India can be divided into three main regions. Some of the highest mountains in the world, the Himilayas, are located to the far north. Then across a section of northern India lies the fertile North India Plain. Hilly plateau regions cover most of the rest of India.

India is also divided into three seasons: hot, cold, and rainy. In the city of Bombay where I live, it is usually quite warm. We do not have any snow. In June we have very heavy rain. It is called the monsoon season, and it rains hard for many days.

Bombay is a busy city. For transportation, people ride bicycles, double-decker buses or trains. India has the fourth largest railroad system in the world. Cars are becoming more common, but gasoline is expensive so it limits their use. We have many vendors who sit on the sidewalk and sell things. It is quite colorful to look up and down the street and see all sorts of things being sold, from fruit to clothing, in these outdoor markets. Usually the streets are crowded with people. Bombay is a very large city with a population of over five million. Since it has a harbor on the Arabian Sea, most of India's imports and exports come through here.

I live in a very large apartment building. My parents and I live in a three-room apartment. I am not allowed to go inside unless I take my shoes off. No one in India goes into a house wearing shoes. My mother wears a *sari*, which is a colorful piece of cloth that is wrapped around the body so that it makes a dress. She also has a red dot on her forehead which means she is married. She loves jewelry, as most women in India do. She wears bracelets on her wrists and ankles and also wears many rings.

Children in India start school at 11 o'clock and go until 5 o'clock. In this way, we have two hours in the morning to do our homework before we go to school. We wear uniforms to school. I wear a white shirt and blue pants. The girls wear white blouses and blue skirts. We have a different teacher for each subject. We also must go to school on Saturdays, but I don't mind, as they are field trip days. We take trips all over Bombay and see many interesting things.

Our favorite holiday in India is Diwali. It is a festival of lights and usually comes in October or November. It begins our New Year. We light oil in clay lamps, shoot off fireworks, and hang bright decorations all around. We get to eat lots of sweets and the boys give the girls presents.

A beautiful tourist attraction which you should see if you come to India is the Taj Mahal. It was built a long time ago by Shah Jahan as a tomb for his wife. You will find many beautiful semi-precious stones decorating the walls. Even the throne he sat on is inlaid with costly jewels. It is one of our great architectural achievements.

India still reflects age-old traditions and customs even under the impact of European influences.

Skill: Reading comprehension

Name _____

Date _____

INTERESTING INDIA

Answer these questions after you have read about India.

1. What are the three main regions of India and where are they located? What important mountain range is found in one of them?

   _____
   _____
   _____

2. How do people travel in Bombay? Which means of transportation would you pick and why?

   _____
   _____
   _____

3. Look up in your dictionary and find the meaning of the word *vendor*. Write a complete sentence using the word.

   _____
   _____
   _____

4. How are the Indian schools like the schools in the United States and how do they differ?

   _____
   _____
   _____

5. Do some research and see if you can find another festival celebrated in India besides Diwali—the festival of lights. Use complete sentences when you write about it.

   _____
   _____
   _____

6. Describe the monsoon season. You may want to do extra reading for more information.

   _____
   _____

Learning Journeys © THE MONKEY SISTERS, INC.

Skill: Researching

INFORMATION ABOUT INDIA

Did you know . . .

- There are 14 main languages in India—the official one is Hindi.

- The population of India is 2½ times that of the United States.

- In India, they wash their clothes in a stream and spread them out to dry in the sun.

- Many people still travel by oxcart in India.

- It took 20,000 laborers and craftsmen 20 years to build the Taj Mahal.

Research more about the famed Taj Mahal and write three facts about this structure and its background.

_____
_____
_____
_____

27. Learning Journeys © THE MONKEY SISTERS, INC.

Skill: Recognizing adjectives

Name _____

SWEET BANANA AND YOGURT SALAD

Yogurt is a necessary ingredient in Indian cooking, especially for those who are vegetarians. Indian yogurt is made from buffalo milk which is richer than cow's milk because of the full fat content. It is thick and sweet-tasting. Yogurt made from skim milk or low fat milk is thinner and more watery; therefore, you should add ¼ cup (60 ml) sour cream to each ¾ cup (180 ml) of yogurt. The Indians use yogurt in a variety of ways: drinks, thickening agents, or mixed with dried fruits and honey, as sauces, in desserts and in salads.

To make your Sweet Banana and Yogurt Salad, follow the recipe below:

2 tbsp. (30 ml) slivered, blanched almonds
2 tbsp. (30 ml) seedless raisins
1½ cups (360 ml) plain yogurt
½ cup (120 ml) sour cream

3-4 tbsp. (45-60 ml) honey
⅛ teaspoon (.62 ml) ground cardomom or grated nutmeg
1 medium-sized ripe banana, peeled and thinly sliced

1. Put almonds and raisins in a small bowl and add about ½ cup (120 ml) boiling water.
   Soak for 15 minutes and drain.
2. Mix the drained almonds and raisins with the yogurt, sour cream, honey and cardomom or nutmeg in a serving bowl. Add the banana slices and gently fold them into the yogurt mixture. Cover and chill thoroughly before serving.

This salad is reserved for special occasions in India.

The introductory paragraph at the top of the page contains several adjectives. Underline each adjective once and underline the noun it describes twice. Then list the adjectives below. Be prepared to discuss how some of the adjectives could be used as nouns in a different context.

_____   _____
_____   _____
_____   _____
_____   _____
_____   _____
_____   _____
_____   _____

Learning Journeys © THE MONKEY SISTERS, INC.

Game directions—India

CHEETAH, CHEETAL!

In India, the Cheetahs, a member of the cat family, were trained to hunt Cheetals, a graceful spotted deer. This is how the following game of tag got its name.

Directions:

You will need seven or more players. Pick one person to be a leader, then divide the other players into two teams. One team will be the Cheetahs and the other team the Cheetals. You will need to mark your playing area with chalk or tape. See diagram below. Players will stand in a line *back to back* about five feet (1.5 m) apart. The leader stands off to one side between the two teams. When the leader calls Chee-ee-ee "tah" or "tal," that team must turn around and chase the other team to his home base line. Anyone tagged before he reaches home is out of the game. The game continues until all the players on one team are out. The other team is then the winner.

You might enjoy playing this game with your friends.

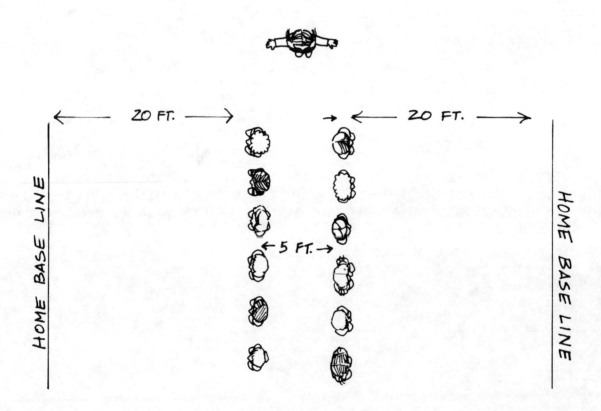

Learning Journeys © THE MONKEY SISTERS, INC.

Skill: 3-place addition—no regrouping   Name _____

Directions: You will need a pair of dice and a partner. Throw the dice and work the problem shown by the total number on the two dice. Have your partner check, then give your partner a turn. Continue until someone finishes all problems correctly. If you throw a number, and you have already worked that problem, you lose a turn and your partner gets his turn.

9.
```
  108
+ 571
-----
```

10.
```
  821
+ 163
-----
```

11.
```
  324
+ 253
-----
```

12.
```
  261
+ 516
-----
```

Doubles
```
  222
+ 222
-----
```

4.
```
  463
+ 234
-----
```

5.
```
  535
+ 463
-----
```

6.
```
  211
+ 345
-----
```

7.
```
  715
+ 142
-----
```

8.
```
  424
+ 364
-----
```

Cheetah!

1.
```
  290
+ 407
-----
```

2.
```
  361
+ 518
-----
```

3.
```
  673
+ 226
-----
```

   Cheetah

1.
```
  232
+ 564
-----
```

2.
```
  156
+ 323
-----
```

3.
```
  344
+ 423
-----
```

4.
```
  143
+ 721
-----
```

5.
```
  259
+ 620
-----
```

6.
```
  525
+ 304
-----
```

7.
```
  478
+ 421
-----
```

8.
```
  245
+ 754
-----
```

9.
```
  657
+ 232
-----
```

10.
```
  717
+ 251
-----
```

11.
```
  306
+ 462
-----
```

12.
```
  465
+ 132
-----
```

Doubles
```
  333
+ 333
-----
```

Learning Journeys © THE MONKEY SISTERS, INC.

'G'DAY' FROM AUSTRALIA

"G'Day." My name is Andrew and I live in Australia. Australia is a continent and the sixth largest country in the world. It is the only country that covers an entire continent. It is about the same size as the United States without Hawaii and Alaska. English is our native language, but we Aussies have our own interesting little phrases or expressions. As an example, "Give it a burl, cobber" really means "Try it, friend."

In the southeast section of Australia, you will find the Australian Alps. Here you will find snow in the winter. Our winter is from June to August. I love to go skiing in the Alps during the winter holidays and so do thousands of other people. About one-third of Australia is desert. Australia has the lowest average annual rainfall of all the continents. Our climate is ideal for playing various sports throughout the year.

Australia is divided into six states plus the Northern Territory which hasn't been granted statehood yet. The six states are: Western Australia, South Australia, Queensland, New South Wales, Victoria and Tasmania.

I live in Melbourne, which is in the state of Victoria. It is an important financial capital. We have a lot of industry such as automotive products, rubber products, agricultural machinery, clothing and textiles. We also ship a lot of fruit to other countries as well.

A popular tourist spot in my state is the penguin parade at Phillip Island. Every night the little Fairy penguins swim ashore. When they get on the beach, they stand up and waddle to their burrows. A spotlight is turned on them when they get on the beach and everyone has fun watching them. The spotlight doesn't seem to bother them in the least.

There is a lot of natural beauty in Australia. In fact, we have the largest rock in the world! It is called the Ayers Rock, and you will find it in the Central Desert. It is famous not only for its size but because it seems to change color from sunrise to sunset.

Australia has become an important source for raw materials that the rest of the world needs. You can find iron ore, bauxite, uranium, nickel, copper, black coal, lead, zinc, oil and natural gas here. Black coal is one raw material that is exported the most.

Most families in Australia live in their own homes. These are called bungalows and have a veranda or a porch. In the big cities, the houses are made out of brick with red tile roofs, and in the suburbs, the houses are made out of wood. Families enjoy doing things together at home. We also entertain our friends by having barbeques or going on picnics.

All children must go to school if they are six years old. When you are 15 you can leave school and go to work, but many continue in school until they are 18 and then go on to college. School starts at either 9:00 or 9:30 in the morning and goes until 3:00 or 3:30 every day. We get five days off at Easter, two weeks off in May, two weeks in September and six weeks from Christmas to February.

The Australians have left their impact on the world in many areas such as medicine, agriculture, aviation, sports and music. Our Joan Sutherland is considered one of the world's finest sopranos.

Even though it is a great distance to get to Australia from North America, I hope you'll visit one day.

Skill: Reading comprehension

Name _____

Date _____

'G'DAY' FROM AUSTRALIA

Answer these questions after you have read about Australia.

1. Australia can claim something no other country can claim. What is it?

   _____
   _____
   _____

2. Australia has winter from June to August instead of from December to March. Look at a globe or a map of the world to see if you can figure out why. Write a brief answer in complete sentences.

   _____
   _____
   _____

3. Draw a map of Australia and label the six states and the Northern Territory. Which state is the smallest? _____

   ┌───┐
   │   │
   │   │
   │   │
   │   │
   └───┘

4. What types of industry are found in Melbourne?

   _____
   _____

5. List the raw materials found in Australia.

   _____
   _____
   _____

6. Joan Sutherland was a famous singer in Australia. Do some research and see if you can find another famous person from Australia that has become world known. Write a paragraph about that person.

   _____
   _____

Learning Journeys © THE MONKEY SISTERS, INC.

Skill: Researching

ALL ABOUT AUSTRALIA

Did you know that . . .

- Australia has a "shopping train" which brings supplies to people in remote areas. They get on the train and shop.

- Sometimes sheep in Australia are rounded-up on a motorbike instead of on a horse.

- About 150 species of mammals in Australia are marsupials. This means they carry their baby in a pouch after they are born.

- In Australia, there are about 700 species of birds.

- A desert with loose stones or gravel is known as a gibber.

Do some reading and list the names of other animals that are associated with Australia.

_____
_____
_____
_____

Skill: Using figurative language

PEACH MELBA

Because of the pleasant climate in Australia, fresh fruit is available year round. Some of their important fruit crops are peaches, apples, pears, bananas, oranges, apricots, plums, and pineapples. They consider the apple their most valuable crop. They use fruit in a variety of ways and quite often for dessert. Peach Melba was created in honor of an Australian opera singer, Dame Nellie Melba.

To make Peach Melba, follow this recipe:

    Peach halves (canned or fresh)
    Individual sponge cake (the kind you use for strawberry shortcake)
    Whipped cream
    Crushed walnuts or almonds
    Strawberry syrup
    Cherries

Place a piece of sponge cake on a small dish. Place a peach half on top with the cut side up. Fill the peach with whipped cream. Pour a teaspoon of strawberry syrup over the whipped cream. Sprinkle with crushed walnuts or almonds and decorate with a cherry on top. You can use vanilla ice cream instead of whipped cream if you prefer.

Write a paragraph using the names of fruits in figurative expressions. Example: "That's just peachy," said Anne. "You know, you're the apple of my eye!" Now let's see what you can come up with!

_____
_____
_____
_____
_____
_____

Art project—Australia

WEAVING

Australia is a major producer and exporter of wool. It leads the world production with its Merino sheep. A shearer can shear between 100-120 sheep in one day. A record was set in 1972 with someone shearing 347 sheep in one day. Shearing is done every year. The sheep could die of the heat, in the hot climate, if they were not sheared. A ram might have 55 pounds of wool sheared off. The wool is then sent to a woolen mill where it is washed and spun into yarn.

Try weaving yourself a bookmark following these directions:

1. Make a loom out of stiff cardboard 8" long and 2½" wide.

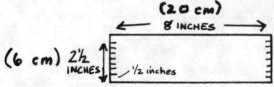

2. Allow ½" (1.25 cm) on each side. Then cut ½" (1.25 cm) notches ¼" (.625 cm) apart on each end.

3. Take some yarn and put it through the first notch. Bring yarn down the front of the cardboard through the first notch at the other end, under the cardboard, and up through the second notch, down the front of the cardboard through the second notch. Now tie the dangling end of yarn to this piece on the under side of the cardboard, come up through the third notch, down the cardboard through the third notch at opposite end, under and up through the fourth, down to the other fourth, under up through the fifth and so on until you finish with the seventh notch. Fasten your yarn on the back side by hooking it through the yarn of the sixth and seventh notches.

4. Use a blunt end needle and thread it with a color yarn you want to weave with. Go over one thread, under one thread, over one thread, under one thread, until you finish the row. To go back, go under first, then over, under, over, etc. Continue with this pattern until you finish your bookmark. Do not pull the yarn too tightly on the edges as you begin each row.

Tie together here.

Fasten here.

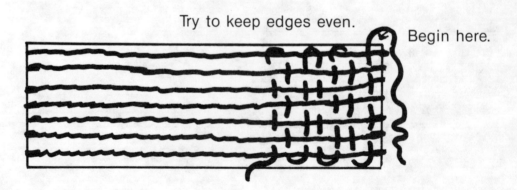

Try to keep edges even.   Begin here.

Skill: 2-place addition with carrying   Name _____

There are 135 million head of sheep in Australia. Find the sheep below with 135 as an answer. Color that sheep tan.

55
+ 28

62
+ 78

49
+ 36

36
+ 46

72
+ 49

13
+ 68

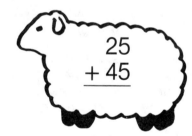
25
+ 45

68
+ 67

76
+ 64

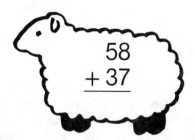
58
+ 37

44
+ 56

27
+ 59

Learning Journeys © THE MONKEY SISTERS, INC.

Skill: Reading

BUENAS DIAS

"Buenos dias!" I am Miguel, and I greet you from our sunny country, Mexico.

We are 1/5th the size of the U.S.A., and we border California, Arizona, New Mexico, and Texas. The two main bodies of water that touch our shores are The Gulf of Mexico and the Pacific Ocean. In the center of Mexico is a high plateau bordered by two volcanic mountain ranges, east and west Sierra Madre.

Mexico is a country that lies north and south, but that doesn't mean that the south is warmer. It is altitude that determines climate. Along the coastlines, climate is tropical. Beaches are lined with palm trees, and people travel from other countries to Mexican resorts such as Acapulco. It is colder in the higher altitudes, but warmed by the sun during the day.

Two peninsulas jut from Mexico's mainland. In the east, the Yucatan Peninsula separates the Caribbean from the Gulf of Mexico. Baja California, a land of mountains and deserts continues south from California into the Pacific.

The capital city is Mexico City. It is located on an island and drained lake where the ancient Aztecs once had a city. Pyramids still stand in some regions to remind us of these people.

I live there in a small house in the suburbs. My father is a teacher, and my mother runs a souvenir shop where tourists like to buy straw bags, straw mats, shawls, jewelry, and art objects. I like to help at the store so that I can practice English. We were settled by the Spanish over 500 years ago. Today, we still speak Spanish as our official language. Our people are descended from the Indians and the Spanish.

Farmers grow corn, cotton, wheat, sugar cane, and coffee. They live in houses built of adobe brick. Straw mats called *petates* are laid on the floor for beds. Colorful flower gardens or hanging pots of flowers cheerfully brighten the house. Each weekday parents travel to the fields to work and children go to school.

More than half the people live in villages or towns. These villages, with pastel painted houses and flower pots on balconies, usually have a plaza in the center and a market place. The plaza is tree lined with park benches where people gather to talk or to listen to music or to celebrate festivals. The city hall and a big church will be found here. The market place is in a large building. Vendors at stalls in the building sell fruits, vegetables, meats, dairy products, clothing and factory-made goods.

Many people move to the larger cities in hopes of opportunities. Guadalajara is a beautiful city with a bright yellow church built over 300 years ago. Merida is built where an ancient Maya city once stood. The Mayas, people who lived long ago, invented a very accurate calendar. Taxco is famous for its silversmiths who make jewelry, and other silver objects. Monterrey is an industrial center.

Our most popular sports are soccer, baseball, and race walking. We like all kinds of music, but tourists always enjoy the mariachi bands. These bands with troubadors dressed in wide brim hats and black suits decorated with silver sequins stroll as they play. Our most famous art are the wall paintings and decorated stone tiles on the outsides of buildings that add color to our cities. Mexican people really enjoy festivals. That's when we enjoy our music and delicious food. The best festival of all is the eve of Independence Day on September 15th. Bells ring! People cheer, and so do I!

Learning Journeys © THE MONKEY SISTERS, INC.

Skill: Reading comprehension

Name _____

Date _____

BUENOS DIAS FROM MEXICO

Answer these questions after you have read about Mexico.

1. Name two of each:
   a. bodies of water that border Mexico
   _____

   b. mountain ranges in Mexico
   _____

   c. peninsulas
   _____

2. What kind of land is Mexico City located on?
   _____
   _____
   _____

3. What are *petates*? Tell how they are used.
   _____
   _____
   _____

4. Describe a village or town. Use complete sentences.
   _____
   _____
   _____

5. Tell something the following cities are known for:
   a. Monterrey _____
   b. Guadalajara _____
   c. Taxco _____

6. Name at least four products grown by Mexican farmers.
   _____   _____
   _____   _____

Learning Journeys © THE MONKEY SISTERS, INC.

Skill: Using a map

MORE ABOUT MEXICO

Did you know that...

- *Sombrero* comes from the word "sombra" which means shade in Spanish.

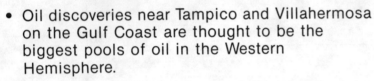

- Oil discoveries near Tampico and Villahermosa on the Gulf Coast are thought to be the biggest pools of oil in the Western Hemisphere.

- *Jai alai*, played by bouncing a ball off a wall with a wicker scoop, is thought to be the fastest game in the world.

- Mexicans love folk songs. The most famous song, *La Cucaracha*, means "The Cockroach."

- No one has counted all of the volcanoes in Mexico. Some are active; some are extinct.

Draw or trace a small map of Mexico. Label the following: Mexico City, Guadalajara, Acapulco, Monterrey, Pacific Ocean, Gulf of Mexico.

Skill: Making a picture dictionary   Name _____

TACOS

Corn has been the staple crop of Mexico for thousands of years. Cornmeal made from crushed corn is shaped into thin round pancakes called *tortillas* and heated.

These soft tortillas are eaten with other foods just as you would eat bread. That's why tortillas are called the "bread" of Mexico!

Some Mexicans make their own and some buy them at the store. You can get tortillas of different varieties in the supermarket. Uncooked tortillas are soft. Tortillas that are fried and crunchy are called *tacos*. These are purchased in packages next to other Mexican foods. Crispy tortilla chips are sold with snack foods such as pretzels and potato chips.

Taco means "snack" in Spanish. But now it is used for a tortilla that is folded in half like a sandwich and fried to a crisp. Then the taco is filled with a variety of foods such as meat, beans, lettuce, tomato, onion, green pepper and cheese. A spicy sauce is then poured on top.

TACO PARTY

Arrange suggested ingredients in separate dishes:
Crisp taco shells
Shredded lettuce
Diced tomatoes
Cooked ground meat mixed with a package of taco spices
Grated cheddar cheese
Bottled taco sauce (mild or hot)

Arrange dishes assembly line style with tacos first, lettuce next, then other ingredients, followed by cheese and sauce. Each person has a plate and walks down the line assembling the taco on his plate as he moves along.

Look up the definition of these Mexican food words. Then create a picture dictionary that includes the word, pronounciation, definition, and illustration. Your dictionary should be bright, colorful, and, of course, in alphabetical order!

burrito	enchilada	guacamole
tostado	tamale	taco
chile	tortilla	nachos

Craft project—Mexico

HOLIDAY PIÑATA

For nine nights before Christmas, special festivities called *posadas* are celebrated. The highlight of the evening is when someone, usually a child, gets to break open the piñata.

The piñata is a papier-mache or hollow clay container. It is decorated with bright tissue paper and usually shaped like a bird, horse, or dog.

Inside the piñata are pieces of candy, small toys, or other gifts. It is tied with a rope that is looped over a bar so that it swings in the air.

A blind-folded child is given a stick, swings at the piñata, and tries to break it. Turns are taken by children until it is broken. When the gifts pour out, the children scramble for the candy and toys.

HORSE-SHAPED PIÑATA

Materials: Newspaper, 14″ (35.5 cm) round balloons, liquid laundry starch, bowl for starch, scissors, colorful tissue paper, construction paper for ears, eyes, and mouth, masking tape, and white glue.

Begin by blowing up balloons and tieing ends. Cut newspaper into 1″ x 6″ (2.5 cm x 15 cm) strips. Roll a double page of newspaper into a tube 12″ (30 cm) long. Make 4 of these for legs. Tape legs onto balloon body with masking tape. Roll another page of newspaper into a 6″ (15 cm) tube for neck. Tape to body. Make a ball from a wad of newspaper to form a head. Tape wad so that it stays together. Tape to neck. You should now have the general shape of a horse.

Dip strips of newspaper into starch. Apply strips so that entire body is covered. Let dry. Apply two more layers. Let dry between layers.

Cut desired colors of tissue into 2½″ (6 cm) strips. Fold lengthwise. Cut into fold to make fringe. Re-fold inside out. Wrap fringe around body and legs. Cover section to be wrapped with glue. Then wrap. Continue until entire horse is covered with fringed tissue. Wrap tissue around yarn dipped in glue for a tail. Glue tail to body. Glue eyes, ears, and mouth to head.

When dry, cut a flap in the side of the body. Fill with treats. Glue flap shut. You are now ready for your PIÑATA PARTY!

Skill: Multiplying by 6, 7, 8, 9 without carrying

Name _____

Help Pedro win the prizes in all the piñatas.

Color one of the numbered candies to show which number you are using for a multiplier. Then write in the number and complete the page.

Learning Journeys © THE MONKEY SISTERS, INC.   42.

NEWS ABOUT NIGERIA

"I putago ula!" That's Ibo for "Good morning!" My name is Wole, and I live in the Federal Republic of Nigeria. Nigeria is on the western coast of Africa. It is more than twice the size of California.

Many people are farmers and live in country regions, where traditional customs are still observed. The city people have adopted modern influences in education, health and commerce. All are proud of their close family ties and traditions.

Nigeria has mangrove swamps and beautiful beaches with palm trees along the Atlantic Ocean. Inland is a rain forest with cocoa and rubber trees. Beyond in the north is a high plateau where groundnuts, cotton and millet are grown. There, for six months of the year, a hot, dusty wind called *harmattan* blows in from the Sahara Desert.

Our rivers are important as "highways" of trade and travel. The most famous is the Niger. It is joined in a Y shape by the Benue. You can see houses on stilts along marshes and creeks. Dugout canoes carry people and goods such as fish or yams.

At Lagos, our capital city, ships from around the world are in the harbor. Oil tankers are loaded here. We are a leading producer of the world's oil. This crowded city mixes modern office buildings, hospitals, supermarkets, museums and TV studios with older buildings and houses. At rush hour there is quite a traffic jam as cars crowd the narrow side streets, and vendors walk along with their wares on their heads.

Our Fulani people are nomads. The families migrate with the seasons, taking their cattle to new pastures. Men wear short, sleeveless garments draped over one shoulder like a Roman toga.

The Yoruba live in large communities. Yoruba farmers will often commute from the cities to their farmlands. They try to become educated and use modern health care. Women wear brightly colored blouses and skirts. Men wear long tunics over loose pants and small caps on their heads.

My family and I are Ibos. We live in a mud-walled house with a rusting corrugated iron roof. My father sells housewares and clothing in an open stall in the marketplace. I like to wear a *dashiki*. It is a loose, colorful shirt that slips over the head.

Nigerians are great artisans. We weave fabrics and beads into beautiful, colorful designs. Decorated pots, gourds, metal, leather and wood carvings are known the world over.

Almost as soon as we can walk and talk, Nigerian children learn to sing and dance. Our dances celebrate important family events, good harvests and festivals. Dancers wear carved wooden masks and dance to the drum, flute and horn. We wear our most colorful clothes, have parades, and feast and dance into the night.

At school I study reading, mathematics and writing. We have school assemblies outside. Sometimes classes are held there, too. Only 30 percent of the children attend school, and most leave by age 12. We use our local language until the third grade. Then everyone uses our official language, English. High school students wear uniforms and have to pass a very hard examination to be admitted.

I hope to have a good job in government, business or teaching some day. My family says that they will help me reach my goal.

Skill: Reading comprehension

Name _____

Date _____

NEWS ABOUT NIGERIA

Answer these questions after you have read about Nigeria.

1. Use your dictionary to find the meaning of mangrove swamp, rain forest and plateau. Use complete sentences to tell how each differs.

   _____
   _____
   _____

2. Name six products of Nigeria. Which is the most important to other nations? Why?

   _____
   _____
   _____

3. What does it mean to say that rivers are important "highways"? Explain in complete sentences.

   _____
   _____
   _____

4. How do the Fulani people and the Yoruba people differ? Why would it be more difficult for the Fulani to become educated?

   _____
   _____
   _____

5. Draw or trace a small map of Africa. Outline where Nigeria is located.

6. Do research to find out more about Nigerian art. Design a face mask or decorated pot in the Nigerian style. Make a picture of your design.

Learning Journeys © THE MONKEY SISTERS, INC.

Skill: Researching

NEWS ABOUT NIGERIA

Did you know . . .

- There are over 200 different groups of people in Nigeria with many different cultures and traditions.

- Lagos means "lakes" in Portuguese and Lagos is a city built on islands.

- 80% of the working people in Nigeria are engaged in agricultural work.

- Camels are used in northern Nigeria to carry people and goods.

- Villagers in the plateau region live in round huts with cone-shaped thatched roofs.

Research more about Nigeria and list two fascinating facts to share with the class. List the book you used.

_____
_____
_____
_____

Learning Journeys © THE MONKEY SISTERS, INC.

Skill: Writing a "how-to" paragraph

GROUNDNUTS

Northern Nigeria is the largest exporter of groundnuts in the world. (You may know them as peanuts!) Only China, the United Sates and India grow more peanuts, but they do not ship as many tons around the world.

At Kano, sacks of peanuts are piled into pyramids while waiting to be shipped by rail to the ports in southern Nigeria. It takes about 9,000 sacks to make one pyramid.

Peanut plants need good soil to grow. The green leaves of the plant grow along the ground, and the peanuts grow on underground stems of the plant.

Peanuts are high in protein and are a good food source. Nigerians use them to make groundnut stew made with a thick groundnut sauce and groundnut cakes fried in oil.

You can make a popular American groundnut food: peanut butter!

Homemade Peanut Butter
　　Roast 2 cups (480 ml) of raw skinned peanuts on a cookie sheet at 300° (154°C) until lightly roasted (about 20 minutes).
　　Place in a blender when cool. Blend, and gradually add ¼ cup (60 ml) of oil. (Peanut oil will add to the taste.) Stop blender to stir large pieces from the sides of the blender. Continue until completely blended.
　　Taste a spoonful or spread on a cracker to taste.

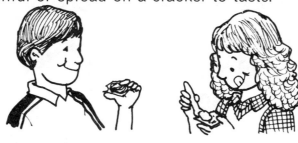

Peanuts are used around the world to make peanut butter, cooking oils and margarine. A famous American of African descent, George Washington Carver, did many experiments with the peanut and the peanut plant. He discovered over 300 uses for this product.

What can you invent with the peanut shell? Use the empty shells left from your peanut roasting. What uses can you think of for these shells? You may crush, break, glue together, or change the shells in any way you wish.

Write about your new uses that will tell others "how-to" do it also. Use complete sentences.

_____
_____
_____
_____

Game directions—Nigeria

NSIKWI

Nigerian children use natural products as well as their imagination for their play.

Clay is shaped into toy dishware and animals. Drums and masks are made from coconut shells to imitate festival dances. Farm goats are pretend horses.

They play tag, Cat and Mouse, Blindman's Bluff, and a bowling game called *Nsikwi* which uses an orange and enough corn cobs for each child playing.

Directions:
Each child playing will need a dry corn cob with the thick end cut flat so that the cob will stand on end. An orange (or rubber ball) is rolled to knock over the corn cobs.

Go outside and sit or kneel in a large circle. Each child stands his cob on end. One child rolls the orange over the ground to knock over another child's corn cob. If he misses, the one nearest the ball gets a turn. If it's a hit, that child gets another turn. Don't worry if the ground is bumpy. That's OK. It just makes the game more challenging!

The child with the last cob standing is the winner and may begin a new round.

This game can be played indoors and made into a multiplication game. Give each child a number from 0 to 12. (Numbers can be repeated.) Assign a number to the ball. When a corn cob is knocked over, the one rolling the ball must state the multiplication fact. For example: 4 times 8 is 32.

Skill: Multiplication facts 6-9

Name _____

Directions: Choose one multiplier. "Hit" corn cob with orange by writing correct answer in each orange.

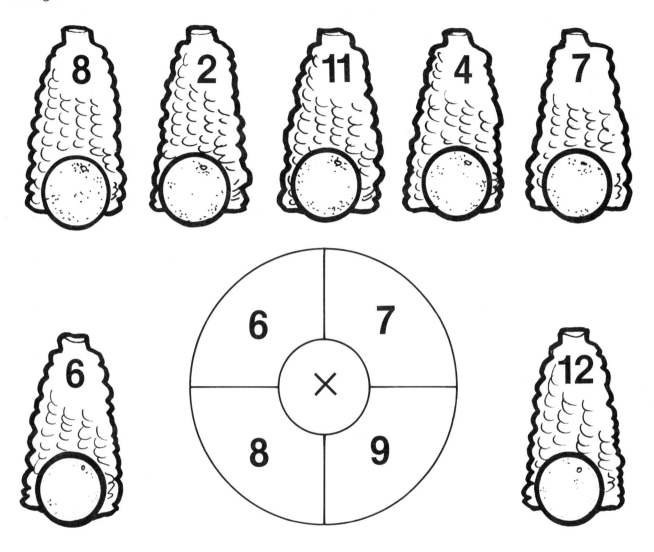

Color in one section of circle to indicate the multiplier you have selected.

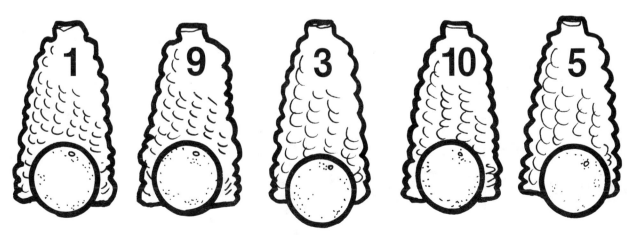

Learning Journeys © THE MONKEY SISTERS, INC.

ANSWER KEY

FRANCE
Page 2:
1. The French gesture with their hands, have expressive faces and dress fashionably.
2. It is one of the most beautiful cities in the world. Famous sights include Notre Dame Cathedral, the Bastille, the Eiffel Tower, the Louvre Museum, the Arch of Triumph and the artists along the river Seine. (Answers will vary.)
3. (Answers will vary.)
4. apples, grapes, olives, lemons, sub-tropical flowers for perfume
5. One would go to a different food store for fruit, bread, groceries, meat, etc. instead of buying all the items in one supermarket.
6. (Answers will vary.)

Page 4:

ap'pe tite	com bi na'tion
tex'tures	to geth'er
in gred'i ents	con ver sa'tion
ap'pe ti zing	e las'tic
ad di'tion al	ar tis'tic
im por'tant	di ag'o nal

Page 6:
68 70 72 74 76 78 80 82
85 87 89 91 93 95 97 99
Red—odd 291 957 717 643 265 243 841
Green—even 176 238 324 750 820 912

GERMANY
Page 8:
1. The northern part of the land is low and flat. In the central and southern parts, the land has rolling hills, plateaus, and mountains.
2. The rivers provide transportation and water power for manufacturing. The Rhine and the Elbe.
3. Wilhelm Roentgen discovered the x-ray and Beethoven was a great musical composer.
4. East Germany is a dictatorship. West Germany is a democracy.
5. (Answers will vary.)
6. A festival is a time of feasting. (Answers will vary.)

Page 12:

18	29	9	9	9	44	13
17	9	35	47	29	9	27
		9	9	9	9	34
		51	18	47	9	26
			36	23	9	16

Pattern loooks like a "9."

ITALY
Page 14:
1. Sicily is located at the toe. Sardinia is located at the knee. The shape of the country closely resembles a boot with a heel.
2. The beaches are covered with pebbles instead of sand.
3. A coliseum is an amphitheater or stadium for public meetings and entertainment.
4. Julius Caesar established the modern calendar and has the month of July named for him.
5. "All roads lead to Rome" means that Rome is a center for learning about history, art, and architecture. It is a cultural center.
6. (Answers will vary.)

Page 18:

87	811	387	810	1019
1001	129	289	189	1001
810	109	775		
810	820	269		
955	410			
840	412			

Mario's violin is the one where "810" is the answer to both problems.

USSR
Page 20:
1. Leningrad 7 AM— Pacific Coast of USSR 5 PM
   Leningrad Noon—Pacific Coast of USSR 10 PM
   Leningrad 4 PM—Pacific Coast of USSR 2 AM
   Leningrad 8 PM—Pacific Coast of USSR 6 AM
2. There are Eskimos, Ukrainians, Georgians, and Armenians among over 100 national groups. Answers will vary as to how they differ.
3. The climate varies from very cold in the Arctic Siberian regions to mild temperatures along the Black Sea. Most of the country has long, cold winters with much snow.
4. Vladimir Lenin was the founder of modern Russia. The city of Leningrad has been named for him.
5. babushka means grandmother
   dacha is a country cottage
6. (Answers will vary.)

Page 24:
1st Award: 5 cm or 50 mm 4 cm or 40 mm
2nd Award: 6 cm or 60 mm 8 cm or 80 mm
3rd Award: 3 cm or 30 mm 10 cm or 100 mm
4th Award: 6 cm or 60 mm 2 cm or 20 mm
5th Award: 4 cm or 40 mm 1 cm or 10 mm
6th Award: 5 cm or 50 mm 7 cm or 70 mm

INDIA
Page 26:
1. Three main regions are the far north, the fertile North India Plain, and the hilly plateau regions. The Himalayas are located to the far north.
2. In Bombay, people travel by bicycle, double-decker bus and train.
3. (Answers will vary.)
4. School in India is from 11 AM to 5 PM and children wear uniforms. There are different teachers for each subject. Saturday is field trip day at school.
5. (Answers will vary.)
6. The monsoon season is in June. It rains very heavily.

Page 30:
Cheetal
1. 697 2. 879 3. 899
4. 697 5. 998 6. 556 7. 857 8. 788
9. 679 10. 984 11. 577 12. 777
Doubles 444

Cheetah
1. 796 2. 479 3. 767
4. 864 5. 879 6. 829 7. 899 8. 999
9. 889 10. 968 11. 768 12. 597
Doubles 666

AUSTRALIA
Page 32:
1. It is the only country that covers an entire continent.
2. Because of the tilt of the earth, the southern hemisphere is further from the sun June-August. Therefore, it gets less direct rays from the sun. (Explanations will vary.)
3. Victoria is the smallest state.

4. automotive products, rubber products, agricultural machinery, clothing and textiles
5. iron ore, bauxite, uranium, nickel, copper, black coal, lead, zinc, oil, and natural gas
6. (Answers will vary.)

Page 36:

83	140	85
82	121	81
70	135	140
95	100	86

MEXICO
Page 38:
1. a. Gulf of Mexico and Pacific Ocean
   b. East and West Sierra Madre
   c. Yucatan Peninsula and Baja California
2. Mexico City is located on an island and drained lake.
3. Petates are straw mats that are laid on the floor for beds.
4. Villages have pastel painted houses with colorful flower pots hanging from balconies. Usually, there is a plaza and market place in the center of town. The plaza is a meeting place as well as a place for celebrations. The city hall and a large church will be located there.
5. Monterrey—is an industrial center.
   Guadalajara—has a large church built over 300 years ago.
   Taxco—is famous for its silversmiths who make jewelry.
6. Mexican farmers grow corn, cotton, wheat, sugar cane and coffee.

Page 39:

Page 42:
(Answers will vary according to multiplier selected.)

NIGERIA
Page 44:
1. (Answers will vary.)
2. Cocoa, rubber, groundnuts, cotton, millet, and oil. Oil is the most important export to other nations.
3. Rivers are used for transportation of products in the same way that highways are used for transportation.
4. The Fulani people migrate and move as seasons change. The Yoruba people live in communities and travel to the farms where they work. It would be harder for the Fulani people to become educated because they move a lot.
5.

6. (Answers will vary.)

Page 48:
(Answers will vary.)